TWICE ACROSS THE PLAINS - 1849 & 1856

WILLIAM J. PLEASANTS

TWICE ACROSS
THE PLAINS

1849 & 1856

William J. Pleasants

YE GALLEON PRESS
FAIRFIELD, WASHINGTON
1981

Library of Congress Cataloging in Publication Data

Pleasants, William J. (William James), 1834-1919
    Twice across the plains, 1849 & 1856.

    Reprint. Originally published: San Francisco: W.N. Brunt, 1906.
    1. Pleasants, William J. (William James), 1834-1919. 2. Overland
journey to the Pacific. 3. West (U.S.)—Description and travel—1848-1860.
4. Pioneers—West (U.S.)—Biography. 5. West (U.S.)—Biography. I.
Title.
F593.P72   1981   978'.02   81-14669
ISBN 0-87770-259-4   AACR2

DEDICATION

To the memory of my noble father, my companion in every privation and pleasure, this little volume is dedicated.

TABLE OF CONTENTS

INTRODUCTION

This little volume, a chronicle of events in the life of the writer during two trips across the plains to California in 1849 and 1856, while in the glamor of his early manhood, is a truthful narrative of his experiences, told in the simplest language at his command, and with a strict avoidance of anything pertaining to exaggeration or distortion.

"THE AUTHOR"

Pleasants' Valley, Sept. 12, 1905.

JOSEPH EDWARD PLEASANTS

Born March 30, 1839 in Warren County, Missouri. Came to California with brother, William, and father, James, in 1849. Settled in Solano County in northern California. Attended a private school in 1856, in home of a family friend, William Wolfskill, in Los Angeles. There he met Mary Refugio Carpenter, a niece of Wolfskill's wife, Maria Magdalena Lugo, married Mary, July 15, 1868, at the Plaza Church in Los Angeles, by Father Francisco Mora. The pictures of Joseph and Mary Refugio were taken at time of their marriage. Mary died January 26, 1888. Joseph Edward Pleasants remarried in 1892. Married Adelina Brown, a school teacher, who died in Orange, California, May 10, 1943.

MARY REFUGIO CARPENTER PLEASANTS

Born July 4, 1845, at Rancho Los Nietos, Los Angeles County, California. Daughter of Lemuel (sometimes Samuel) Carpenter, mountain man, and his second wife, Maria de Los Angeles Dominquez. Carpenter bought Rancho Santa Gertrudes from Josefa Cota Nieto Grant. Carpenter lost the Rancho for debt in 1859 and took his own life. Ex. Gov. of California, John G. Downey, was the buyer of the Rancho, and because of public sympathy for the widow, gave her the original homestead and one hundred acres.

CHAPTER ONE

OR THE SAKE OF those of my own blood who may some day read these pages, it might be well in the beginning to refer briefly to the foundation of our family in America.

Its history on these shores commences with the early settlement of the State of Virginia, when, in the year 1668, one John Pleasants, a Quaker, came over from Norwich, England, and located at "Curles," a little town on the James River, not far from where the City of Richmond now stands.

To trace the family name through the successive generations of that early period would require much space, and therefore be an injustice to the general reader; hence I will omit the data bearing thereon and begin again at the year 1806, at which time Edward Pleasants, my grandfather, emigrated with his family from Goochland county, Virginia, to Lincoln county, Kentucky.

It was here that my father, James M. Pleasants was born on the 29th day of April, 1809. August 8, 1833, he was united in marriage to Miss Lydia Mason, and two years later, being desirous of going further West, he moved to Missouri, where he located. Here in the peaceful pursuits of a farmer, he reared his family.

In January, 1849, the news of the discovery of gold in California reached Western Missouri, where our home was and spread with great rapidity throughout the entire region and soon nothing else was talked of in country or in town. Farmers, merchants, mechanics, lawyers, and even ministers of the gospel, fascinated by the wonderful stories of vast wealth uncovered by the miner's pick in the far-off land bordering on the mighty Pacific, shared in the general excitement, and so much interest was manifested in the subject of emigration to that distant region that the question now was, not who would go to California, but, rather, who would stay at home.

11

Some, however, from the very beginning, argued that it was foolish and foolhardy to leave a prosperous community in order to embark upon a venture that was not only dangerous in the extreme, but was at best doubtful of results and liable to prove in the end disastrous to all concerned. And, indeed, it did take courage to tear one's self away from a happy home, the delightful society of relatives and friends, and plunge into the depths of a vast, almost unknown wilderness, roamed by savage wild beasts and still more savage men. Now and then, too, there would drift in from that distant region tales of desperate attacks on lonely caravans and of the ruthless slaughter of men, women and little innocent children by the hordes of Indians that infested the wild lands lying between the outposts of civilization and the far away land of gold.

After the first wave of wild excitement and enthusiasm had died away and men had time to reflect a season on the perils of such an undertaking, and could in a measure realize the hardships necessarily incident to a prolonged journey through a country without roads or other appurtenances of civilization, many of those who had been among the first to volunteer for the trip rid themselves of the outfits already purchased and signified their intention to remain at home.

But the more adventurous spirits, those whose names were afterward linked imperishably with the history of progress and vigor and manhood of the West, were only too anxious to encounter and grapple with the dangers lurking everywhere along the lonely trail leading to the distant land of promise.

My father, then in the full strength of his manhood, positive and daring in his nature, and with an abiding faith in the future greatness of the vast empire stretching out in limitless grandeur towards the West, had long been possessed of a desire to turn away from the conjested district where we then lived, and be among the first to seek a home and found an estate somewhere beyond the boundaries of the Rocky Mountains.

My mother had passed away, leaving him with six children, and with the family tie thus broken and the news from California coming, as it did at this time, just a few months after our terrible bereavement, my father determined to carry out his long cherished plans and at once began his preparations for the journey westward. He had two good light wagons that he had made with his own hands, which were well adapted for a trip across the plains. Two friends, good, strong men, John and David Burris, volunteered

12

to accompany him. These, with my father, James M. Pleasants; my brother J.E. Pleasants; and myself made up the five that composed our party. Our outfit, altogether, consisted of the two wagons, five yoke of good well-broken work oxen, one saddle mule, sufficient clothing to last us eighteen months, for that was the length of time we expected to be gone, and a six months' supply of provisions, principally bacon, flour, sugar and coffee. Actual experience afterwards taught us that we had made a mistake in not having with us more dried fruit, rice and beans. Each of us was armed with a good gun and there were several pistols also in the party. To provide against rainy days our wagons were covered with heavy rain-proof canvas, stretched on bows. Altogether we did our best to provide everything necessary for so serious an undertaking, but after all our studied forethought in regard to the matter, the result was far from perfect, but this realization came too late to remedy. While we were thus bending every energy to hasten the preparations for our departure, scores of others were doing the same, for, of course, our little outfit was to be one of many that would go to make up a large caravan, banded together for company and mutual protection.

It was at this time, when we were almost ready to gather at the starting point that the weak-kneed ones began to falter, and, indeed, it was not a pleasant picture to conjure up, this two-thousand-mile journey over desolate wind-swept plains, high snow-capped mountains, burning, waterless deserts, deep rivers to swim, treacherous quicksands, and, more terrible than all else, the deadly rifle and keen scalping knife of the lurking savage.

It took courage to face all these, and, at last, when the day to start did arrive and the long wagon train moved slowly away in the path of the setting sun, leaving mothers, wives, sisters and sweethearts and civilization behind, there was not one single coward accompanying it.

It was on the sixth day of May, 1849, at 9 o'clock A.M., that our little party started away from the beautiful town of Pleasant Hill in Cass county, Missouri, bound for the Lone Elm, the place agreed upon as the rendezvous for all those who had enlisted for the expedition. On this beautiful May morning, when all nature seemed in her loveliest mood, we solemnly bade good-by to friends and loved ones and drew away just as the village school bell was tolling.

The Lone Elm, for which we now headed, was a solitary tree standing far out in the wide prairie, some fifty miles from the nearest settlement, and we were about two days in reaching it. At this point, when all had been

13

The Start from the Lone Elm

14

accounted for, we proceeded to organize and elect officers. James Hamilton was chosen captain and John Lane, wagon-master. Written regulations for the government of the expedition were drawn up and adopted, and then with hearts filled with high hopes for the future, one hundred and twenty souls, all told, with grim determination, set their faces against the West and moved slowly along over the flower-bespangled prairie. There were but few women in the party and only four boys, two of the latter being my brother and myself. All these people hailed from the western counties of Missouri with the exception of a limited number, who had come from other portions of the country in order to join our company. The total number of wagons comprising the train was thirty-three.

Just here I will say that the main object in view with nearly all of us on this expedition, was to dig gold, and we really expected to obtain from that source within a few months sufficient riches to return home independent. But some of those among us had friends who were already in California. A brother of Mr. Lyons was at Vacaville; William Hopper had a cousin, Charles Hopper, living in Napa Valley, near Yountville. These gentlemen may have had other things in view, but as I have before stated, with the majority of us it was simply to enrich ourselves in the gold fields that we went to California, though of course a love of adventure and exploration in a new country, fascinating themes to most men, may have greatly influenced us also. Our expectations along these latter lines were in the end fully realized, but the collection of wealth that we had dreamed of never materialized, though a few, just seven, I believe, did succeed in reaching the five thousand dollar mark and over in the diggings. One could find gold almost anywhere, but to find it in paying quantities was another question. But it is my intention to dwell more fully on this subject further on.

We will now return to the Lone Elm, that giant sentinel standing on the outskirts of civilization. Here, awaiting the signal to begin that memorable journey through the wilderness, stood the little band of devoted men and women. This day meant much to them. Would a kindly fate smile and bring them weal or a frowning destiny fill their lives with woe. Who could tell? Finally, all being ready for the start, John Lane, the master of transportation, gave the word. One after another the wagons fell into line and soon in a straight string over half a mile in length the Pleasant Hill Train, so named in memory of the town that bade us Godspeed when we departed from it, moved slowly westward across the trackless prairie.

15

During the first few days out the weather was delightfully calm and beautiful, the atmosphere pure and exhilarating, and the spirits of all our company correspondingly cheerful. With an abundance of good feed for our stock, the luxuriant prairie grass being from eight to twelve inches high, and the whole landscape, as far as the eye could reach in every direction, bearing the appearance of a rich meadow, everything did, indeed, seem propitious and the future full of promise. But how little we can reck of the morrow. At that very moment there was lying concealed in our midst a grim monster, only awaiting a favorable opportunity to seize upon and cut down without a moment's warning some of the bravest and noblest of our little band.

Since 1848 a terrible epidemic of cholera had been raging in the States of the Union, claiming its victims by the thousands, but we naturally supposed that being so far away from the busy haunts of men, and the noisome influences of over-crowded cities, living clean lives, our own hearts close to the great heart of nature and beating in unison with it, that we would be beyond the reach of the destroyer. But from this dream there must soon come a rude awakening. Within a few days we were to feel the sharpness of the monster's sting and leave in its merciless grasp comrades that we had learned to love and honor.

From five to fifteen miles is the varying record of our daily travel. The distance made is, of course, dependent on many things; whether the ground happens to be soft or firm, the country hilly or level. And the crossing of water courses is a mighty factor in the matter. Occasionally there are low flat places where the soil is so miry, hours are consumed in going a short distance. During the latter half of May and the early days of June, there is much cloudy, rainy weather, making travel very difficult, the swollen streams being among the worst features to overcome. In passing through that section of country that is now the State of Kansas, we met with a great many Indians belonging to different tribes, but they were invariably friendly and gave us no trouble. In fact, more than once we found them valuable allies when it became necessary to cross some of the larger streams. Our stock could be made to swim to the other side, but the wagons must be ferried over. So, procuring from our savage friends about eight canoes, we would lash them firmly together, side by side, and then across the whole lay strong poles about one foot apart. These were fastened securely to the canoes with inch-wide buffalo thongs, and with this hastily improvised structure the wagons would be safely ferried across the widest and most turbulent streams. The Kansas

16

The Burial of John Lane

River, then called the Kaw and the North Platte, were both crossed in this manner.

Our train was now some two hundred miles away from its starting point. Buffalo and antelope could be seen by the thousands. Many of my comrades were old hunters and unerring shots with the rifle, hence the camp was always well supplied with fresh buffalo hump, antelope steak, and now and then for a change buffalo marrow bone would be served. These were great times for the younger sportsmen. The duty of driving wagons every other day devolved upon them, but their leisure moments were spent in the chase, and their share in the exciting sport was by no means limited. And now, just here in the midst of these pleasures, occurred an event that cast a pall of gloom over all our company. At about 2 o'clock one afternoon John Lane, our wagon-master, a man who, by the splendid attributes of his character, had greatly endeared himself to all those who accompanied the expedition, was suddenly stricken with cholera. The train was immediately brought to a standstill, wagons corralled, oxen unyoked and turned out to graze and every effort made to relieve the sick man. All the simple remedies known to us were applied, but these endeavors came to naught, for our friend grew gradually worse and after suffering intensely, died in the middle of the forenoon next day. Being entirely without those articles usually employed in such cases we prepared his body for burial by simply wrapping about it the nicest blanket we possessed and then laid him to rest on the north bank of a beautiful river, known as the Little Blue. Erecting above his grave head and foot boards bearing a suitable inscription, we turned sorrowfully away from the lonely mound, leaving our friend to the care of Him who said, "Lo,' even in the midst of the valley and the shadow of death I am with you."

18

CHAPTER TWO

HERE WAS little joy among us as the journey that had been so sadly interrupted was resumed. Knowing now that we carried with us the fatal germs of cholera, and too familiar with the history of the dread disease to believe that it would be satisfied with a single victim and would henceforth leave us in peace, each individual among us wondered in his own mind who would be the next unfortunate. And the question did not long remain unanswered.

The beautiful weather that had been favoring us for some time was now a thing of the past, and it rained almost continuously. Another case of cholera now developed, followed by still others. We fought the enemy hard and were successful in effecting cures in three-fourths of the cases, but found it impossible to save all, and Julius Wright, a most estimable young man, the only son of a poor widow, succumbed to the disease. His mother had encouraged him when he had expressed a desire to take the trip to California, believing in her loving heart that it would redound to his benefit, and this was the end of the dream. His last thoughts were of her, and the sweet message he left for the poor woman must have been as balm to her bleeding heart.

We have now reached the south bank of the Platte River. All the streams are greatly swollen from the heavy rains which still continue almost uninterruptedly; the mud is deep and traveling very difficult. There is an abundance of fine feed for our stock, however, which is some compensation for the troubles that so continuously harass us. The Platte River Valley is about one mile wide, low and level, and covered with a rank growth of grasses. In order to find a suitable camping place for the night, we pull away from the trail and direct our course towards a spot some half a mile distant. Several of our party, among whom was William Hensley, an exceedingly lively, jovial, good-natured fellow, were walking along together toward the point mentioned, when he laughingly said, "Boys, this would be a dreadful place to be buried in. I should hate to die and be laid away in this low, muddy flat." Somehow the remark struck us as being significant, for we could not help living in constant dread of that terrible unseen foe that was dogging our footsteps across the continent and would not be shaken off. Poor Hensley was the next victim. At about 8 o'clock that evening he was seized with the awful cramping that is the initial stage of the disease, and after suffering all night in fearful agony, died at 8 o'clock the next morning, and he was buried in that same muddy valley. And so the irony of fate decreed that the program, the mere thought of which a few hours before had occasioned him the greatest horror, should be carried out to the very letter. Unnerved and discouraged by the death of their friend, Hensley's messmates, Middleton Story and Emanuel, a negro, concluded to secede from the company and return to their homes, now about six hundred miles away. We argued and plead with the two men for an hour or two in our endeavors to dissuade them from risking so hazardous an undertaking, for it seemed to us that it was far more dangerous for so small a party to go back than it was for all to go forward, but they refused to reconsider their determination and return home: so, bidding them good-by, we pushed on, leaving their lone wagon standing, still surrounded by a few faithful friends, yet pleading with them to remain. Finally the negro, laying the whip in Story's hand, said: "Mid, do as you please, but no matter what your decision may be, remember I am with you. If you return, I will go back also; stay, and you will still find me by your side." Story gave the off ox a sharp blow with the whip and the two animals, anxious to rejoin the herd that had so long been their constant companions, swung into line as they had done so often before, and our two comrades were once more headed for California, and from that time on until we reached our journey's end swerved neither to the right nor left, but kept their eyes towards the setting sun, and followed where it led until they saw it sink to rest in the mighty waters of the Pacific Ocean.

The weather is once more clear and beautiful and the health of our people much improved as our train winds up the south side of the Platte River, which at this point is about one mile wide. The current is swift and the water cold and muddy. Very shallow, it can be forded almost anywhere, provided one keeps moving after having once entered the stream. If while crossing, a wagon is brought to a standstill, the swift current washes the sand from underneath the wheels and causes it to sink so quickly that a very limited stop may cause the loss of a valuable wagon and cargo.

The country lying south of the Platte, with its table lands and low rolling hills, is a favorite feeding ground of the buffalo. Here they may be seen in countless thousands peacefully grazing. In going to the river for water these animals seem to move in a straight line, one after the other, like an army of men on the march. Migrating from the far South, where they have passed the winter, the vast herds move slowly towards the distant river to quench their thirst in its sweet waters. The first one leaves a slightly marked trail, another naturally follows this, and a third cuts it a little deeper still, until finally the hundreds and maybe thousands that have passed that way have left a straight, narrow, sunken path, several inches perhaps below the natural surface. Then the rain comes. The land having a gentle incline towards the river, the water flows down these little ditches, carrying with it the loose trampled soil at the bottom, leaving gullys from three to six inches in depth. These are three or four feet apart, extend far back towards the uplands, and are a feature of the land lying contiguous to the river for miles up and down the stream. Sometimes we are compelled to travel long distances along this uneven surface and the incidental and continuous bumping and jolting of the wagons is disagreeable in the extreme.

Since the good-by to civilization on that beautiful morning in May last, though many troubles have beset our pathway and we have drunk of the bitter dregs of misfortune, the expedition has much to be thankful for. The cattle and other stock that started with us are still in good condition and doing excellent service, which is no small item when one stops to consider the terrible consequences that might result should the men and women of our party be suddenly deprived of the means of transportation here in the midst of the vast wilderness. It would not be possible to proceed on foot and carry with us a supply of food and other necessaries sufficient to complete the journey. At all hazards the great mountain ranges lying to the westward must be crossed before the snow flies, otherwise our situation would be pitiful indeed. Hence all along, being thoroughly alive to these dangerous possibilities, we have from the very beginning taken every precaution to safeguard these vital interests. If during the day we lay by for any purpose, our oxen, on being unyoked, are immediately taken in charge by two men, who stay with and watch them closely to see that they do not scatter or are not stampeded by Indians or other causes. After darkness sets in, these are relieved by four fresh men, who remain until the hour of nine, when four others take their places until midnight, and this routine continues while darkness lasts. So there is not a moment of the day or night when our animals are not under the strictest surveillance.

Continuing up the south side of the Platte for a week or two, we pass the point where the North Platte empties into it on the north side. About twenty or thirty miles above the junction of these two streams Thomas Moore died of the cholera.

Just here we make preparations to cross to the other side. This looks to be a difficult undertaking, for the water is nearly if not quite one mile wide. First men on horseback are sent across to gauge the depth of the channel and choose a route that presents the least difficulties in the way of our crossing. These matters being settled to our satisfaction, we raise all goods as high up as possible in the wagons to keep them dry, and begin the passage with one horseman in front to guide the teams and two on either side to see that they are kept in place and continue moving. Much time is consumed in making this portage, but finally, barring a few slight accidents, all are safe on the other side.

There is one more cholera victim here — an old man named Kearns. He lies buried on a bluff north of the river.

Since our start and up to the present time our course has been a little north of a due west line. The general character of the country has been low, rolling ridges and table lands; the soil, rich, black and covered thickly with nutritious grasses. There has been no timber to speak of except along the streams and that only a scrubby kind of cottonwood and willow.

Wood for campfires has at times been unobtainable, but usually we have had a small quantity of dry wood with us, carried for just such emergencies from some previous camping grounds where there was an abundance. If in a portion of the country where there was danger to be apprehended from Indians, we have avoided stopping for the night near streams where there was timber and brush, not caring to take any chances along those lines. Frequently, however, when there was every indication that we would be free from molestation, we would seek the little creeks for our bivouacs, for along these there is always more or less brushwood, good material for small fires. Our people had just crossed over the South Platte River when I digressed for a moment to speak of the general topography of the country through which the expedition had traveled up to date. I will now resume the thread of my narrative.

Crossing the South Platte

Leaving the river at this point of crossing we move in a northwesterly direction through a rolling hill-country and after making fifty miles or so reach the North Platte, a fine stream about, I should judge, four hundred and fifty yards broad. Deep, swift and icy cold, its waters are dangerous, but we cross it without serious mishap, many of us swimming it over and over again while engaged in rendering assistance to the teams and wagons.

Following up this stream, our course lying parallel with it, we travel for several days sometimes quite near it banks and then again miles away. This is the Black Hills country, destined in the far future to be a wonderful gold-producing region.

And now listen: It is noon, and while stopping for a brief interval to rest and breathe our cattle, we are hailed by a man garbed in the dress of a frontiersman, fully armed and speaking with a broken accent. He boldly approaches our train and after questioning us closely concerning our pilgrimage, destination, etc., does not hesitate in turn to fully set forth the reasons for his own presence in this secluded locality. A French Canadian, he claims that his occupation is that of a trapper. In age, somewhere near thirty years, wiry in build, quick and elastic in his movements, he is a splendid type of the class of men roaming the wilds of the great West in the interests of that giant organization known to the world as the Hudson Bay Fur Company. This stranger is friendly and confiding, his manners agreeable and his personality rather unusually attractive. "Why undertake the hazardous journey to California in quest of gold," said he, "when right here in these hills, almost within a stone's throw of where you are now standing, and to be had almost for the asking, is virgin gold in quantities that would stagger the dreams of avarice. Squaws in the Indian villages beyond the river dig it in quantities and shape it into rude ornaments for their own personal adornment. If you doubt these assertions of a stranger follow where I lead and verify with your own eyes the truth of my statements."

Although our interest is wonderfully aroused by his story, we are chary of trusting this man. His occupation is that of a trapper, and traps are sometimes set for men as well as for beasts. And it behooves us to be wary, for in these days there are rife startling stories, told around the campfires in the gloaming, of little bands of emigrants while wending their way along the lonely trails, lulled into fancied security by a long season of peaceful surroundings, relaxing for a time the seemingly useless stern vigilance so long practiced that it had grown irksome, being enticed by the fascinating story of

The French Trapper

some plausible stranger, into a deadly ambush, the fortunate ones slain, while the unfortunates, those made captive, are led away to be put to a lingering death by every species of torture that a devilish ingenuity could suggest. Tortures, the details of which cannot be put into print, but so horrible in their cruelty that could they be given to the reader it would almost freeze the very blood in his veins.

But gold seeking was our business and we finally concluded to investigate the trapper's story. With this end in view, Captain Hamilton gave orders for the train to lay by for a day or two, and the next morning ten men, well armed and mounted, set out, guided by our trapper friend. A cold rain was falling and as the river had to be crossed to reach the Indian village lying on its northern side, the men, not relishing a swim in the icy waters in such weather, abandoned the undertaking and returned to camp. This closed the incident, but it afterwards occurred to me that the man was sincere in his intentions and truthful in his story, for since that time these same Black Hills have added many millions of dollars to the world's supply of the precious metal.

CHAPTER THREE

E ARE AGAIN ON THE MOVE. The country gradually becomes more broken and hilly. High mountain peaks loom up in the distance, the nights grow colder, grass less plentiful and good camping places hard to find. Laramie Fork is the next stream we meet with. It is swift, clear, cold and about twenty yards wide at the point where we forded it. Here is situated Fort Laramie, where a garrison is stationed to keep the Indians in check and protect emigrants. After several days travel up the south bank of the North Platte we cross over to the other side. This crossing was the most difficult and dangerous and we consumed more time in making it than any similar event yet encountered. A crude ferryboat constructed of Indian canoes, in the manner previously described in this narrative, was the means by which we reached the opposite shore. It was a slow process and we were two days in completing it. This boat was the property of a trapper, was propelled with oars and paddles and but one wagon could be carried at a time. The river was about three hundred yards wide and so swift was the current it was necessary to pull or tow the miserable craft two or three hundred yards up the stream before starting, and then it would land about the same distance down stream on the opposite side. If I remember rightly we paid five dollars each for the ferrying over of the wagons. The cattle were made to swim across, an operation both tiresome and dangerous to the men who had charge of it, and requiring patience and skill also. A man on horseback would ride into the stream and the cattle were driven in immediately after him. For a time they would follow him in good order, maybe cover fifty or a hundred yards, and then, being partially blinded by the reflection of the sun in the water or for some other cause, they would turn from the straight course previously pursued and begin to swim in a circle. The men on horses who were behind driving them would frequently get caught in the midst of this struggling mass

27

of heads and horns, their efforts powerless to prevent the whole herd from finally returning to the same side from which they had recently departed, some landing maybe a half mile below the starting point. This monotonous proceeding was repeated many times before we finally had the satisfaction of seeing all of them on the opposite shore.

Here, after conferring together on the subject, we conclude to divide our company into three divisions or separate trains, not because of any ill feeling or misunderstanding among us, but for the simple reason that we have now reached a section of country where stock feed is becoming scarcer all the time and our cattle losing flesh and strength, and the smaller the number of cattle the farther the feed will go. So we part company. About due west is now our course, and ere long we are among the sage brush, where we find good dry grass along the streams. For some days more our train is following up the shore of the North Platte, and during this time we cross several small streams running from the North and emptying into the main or larger river. One of these tributaries is known as Grease Creek, named evidently from the great quantity of grease wood brush growing along its banks. This peculiar growth will burn as if covered with grease, hence its name.

It is quite evident we are now out of the ranges of the buffalo. I do not recall ever having seen a single one after crossing the North Platte. But antelope are very numerous and black tail deer plentiful enough in the quaking asp timber growing about the heads of small streams. There are many grouse and numbers of a species of prairie chicken. We pull away from the North Platte along in the latter part of July and pass Independence Rock, so named by John C. Fremont, who celebrated the Fourth of July, 1846, at that point. This huge rock, which is forty or fifty feet high, is an interesting landmark, standing, as it does, with none of its kind about it. I climbed up onto its top and wrote my name there.

A few miles west of Independence Rock flows the Sweetwater, a stream two hundred miles or more in length, but shallow and easily forded anywhere. This river flows through a wonderful canyon, known as the Devil's Gate. It is a narrow gorge, only about sixty feet wide, and probably three miles in length. The walls on either side are of solid rock, about three hundred feet high and perpendicular. Standing at the bottom and looking upward, the crevice at the top, owing to its distance from the observer, seems so narrow that a man could easily leap across it, and it would appear a certain

28

antelope thought so, too, for I found the body of one lying dead on the rocks at the bottom, having evidently fallen from the very top. A human being would find it impossible to go through this canyon on account of the rough character of its interior, and it is likely that no one but the slippery gentleman in whose honor it was named ever succeeded in making the trip alive. But there is no authentic record in existence showing that even he ever attempted it. We followed up the Sweetwater until we had traversed its entire length. It took us about twenty-five days to do this, and nothing worth mentioning occurred during that time. The weather is now dry and beautiful and the health of our people good. On or about the fifteenth day of August we are at the summit of the Rocky Mountains, that backbone of the American continent. Since leaving the Lone Elm, our starting point, some ninety days ago, our course has been up hill; not noticeable, of course, the greater part of the way, so gradual has been the incline, but up hill just the same. In all this time no word from home has reached us, neither has there been an opportunity for us to send a message back, hence the anxiety of our loved ones must be extreme. They know that a hundred dangers threaten us to one that threatens them. Each morning they arise with the hope that the day may bring from out of the great silent wilderness some word, some reassuring message concerning us. But the days lengthen into weeks and weeks into months, and still that awful silence and uncertainty, and hope deferred that maketh the heart sick.

Up to date we have lost by death altogether five comrades, four from cholera and one, Mr. David Rice, died of fever. We have seen thousands of Indians, traded ponies with them, and exchanged flashy red blankets for buffalo robes, sometimes receiving several good robes for one blanket. Captain Hamilton brought with him two race horses. While passing through the Sioux Indian country, these natives being good judges of horseflesh, were very anxious to get possession of them, offering six ponies for one, a beautiful sorrel. This being refused, they followed in our wake for several days, still hoping to secure the prize. Each morning they would examine the racer more closely, and then seemingly having found some new and heretofore undiscovered point of merit in the animal, would raise their previous offer a pony or two, always without success. This was kept up for some days, when finally a whole herd of ponies was proffered, but the captain was obdurate, so the would-be traders departed in disgust.

The Indians of the plains have many customs that are strange to the white man, among these being their manner of disposing of their dead. Selecting a tall tree with spreading branches, they lay along these small poles, which, being securely fastened in their places, form a rude level platform. The body, after being wrapped in a buffalo robe, though sometimes a blanket is used for the purpose, is laid at full length upon the structure. And then, in order that the journey to the happy hunting grounds may be made in comfort, a small vessel of water is placed upon the breast of the departed brave and left there. This ends the ceremony. Sometimes several of these aerial graves may be seen in the same tree. Some of our boys climbed up to where these grewsome relics were deposited to view them more closely, but as for myself I was as close to them as I desired to be while standing on the ground.

But from this retrospect we will return to our train, left standing on the summit of the Rocky Mountains.

This range is sometimes called the Great Divide, for it divides the watersheds of the continent. East of it all the waters flow into the Atlantic Ocean; west, they find their way into the Pacific. Where we now are the mountains are comparatively low, for this is the South Pass, one of the few points along the Rocky Mountain system where it is possible for wagons to cross without encountering almost insurmountable difficulties. North and South of us as far as the eye can reach are the higher portions of the range, with great peaks here and there, many of them covered with perpetual snow. Viewed through the clear, exhilarating atmosphere of this high altitude the panorama spread out before us is wonderful and grand, and we are inspired with new hope and new zeal, as we realize our journey is half finished and over there where this water at our feet is flowing is California, the land of our dreams. With these pleasant thoughts uppermost in our minds, we make camp for awhile, just at noon, on the brink of a gigantic spring of ice water, known as the Pacific Spring, for its waters flowing westward empty themselves into the great ocean of that name. Our course from here is due west, through rolling hills, covered with sage brush. The soil of this section is poor and sandy. There is no timber along our line of travel, though far away on the higher elevations we can see some.

Within three days we are at Green River, whose waters are very cold, coming as they do from the melting snows on the peaks. It is only about twenty yards wide, and being fordable, we are soon on the other side. Next we climbed the Bear River Mountains, a most difficult feat, as they are very

The South Pass

steep and rough and broken, but when at last the summit was gained, we saw far below one of the loveliest of landscapes, the beautiful Bear River valley.

Lingering for a moment to admire the superb picture lying thus spread out before us, we begin the descent. Here are to be seen the old trails used years before by emigrants bound for Oregon. It is said of these early travelers that they were compelled to use ropes in many places to let their wagons down into the valley, so rough and steep was the way, but our party, steering clear of these old roadways, selected a new route and reached the bottom without having to resort to any such devices.

About forty yards wide and very deep, the river here runs northwest, but further on turns west, then southwest, and finally empties its sparkling waters into the Great Salt Lake.

For about eighty miles this stream runs through a beautiful valley from one-half a mile to a mile in width, and we followed along its banks for that distance. I cannot bid farewell to this lovely vale without expressing my admiration for its many attractive features. The soil is rich, in fact the best we have seen since leaving the lower Platte, and throughout its entire length there was fine feed for our stock. Great fields of wild flax of an excellent quality, the bushes ranging from three to five feet in height, were seen growing here and there, sometimes as much as fifty acres in one body. Small game, such as grouse, abounded and fish were so plentiful in the river that many times we feasted upon them. There was something homelike about everything pertaining to this section, and it reminded me of the country where my early youth was nurtured.

But I turn away with regret from this lovely scene, as yet untouched by the vandal hand of civilization, feeling that the memory of its wondrous beauty would linger with me and grow more mellow with time as I journeyed slowly down the path of years.

CHAPTER FOUR

CHAPTER IV.

UR COURSE NOW LAY between old Fort Hall and Great Salt Lake, about an equal distance from each, through hills covered with sage brush. 'Tis a desert country that we are now in, and feed for our animals is so scarce that they begin to lose flesh rapidly. Owing to the loose sandy nature of the soil, the pulling is hard, and it is not possible to make much more than a snail's pace.

As we proceed the general aspect of the country becomes more and more forbidding, and we begin to anticipate trouble ahead, but there are no signs of discouragement to be seen among the members of the expedition. From the very beginning all have had a supreme confidence in our ability to reach the destination aimed at. What others had done we believed we could do, and so, making merry over the troubles that so constantly beset our pathway, we pursue in a cheerful mood the tenor of our way.

I wish to say here that if one desires to know the true nature of a man, the nature from which certain hard conditions will strip the mask of sham and pretension and lay bare from under its false surface the individual as he really is, journey with him day after day for months in a lonely land teeming with difficulties that only patience and perseverance can overcome. Watch his daily life that is no longer influenced by the artificial conditions and requirements of tne vast network of society that he has left far behind him, his only associates now being the little handful of men and women who are subject to the same stern conditions that apply to him — and note the result. If he be brave and tender and true, it is not his own burning thirst out there amid the hot sands of the desert that is urging him on in a frantic search for water. Oh, no; he can stand his own suffering yet for awhile, thank God, but the pitiful cry of the poor fever-stricken woman in the wagon back there at the rear end of the train is more than his manly heart can bear.

If he be a coward — oh, well, no need to dwell on his qualities; the very stones along the roadside know and recognize him as he passes by. And so uncharitableness, selfishness, untruthfulness, all in turn and in good season show their hideous heads if they exist in the individual. No double life here, no pretension. A man is, simply, what he is, nothing more, nothing less.

A few days more in a flat, sandy region brings us to a hilly section of country through which flows a stream called Goose Creek, a tributary of Snake River, which we follow for a day or two, then turn from it, bending our course toward the southwest, and presently reach Thousand Spring valley, an uninviting spot lying flat and marshy between low hills. This valley is about one-half a mile wide and several miles long, and takes its name from the great number of springs or overflowing wells found within its borders, many of them being as much as twenty feet in diameter, with no visible bottoms. The country through which we are now traveling is all in all the least attractive of any yet met with, being too poor for even Indians to live in. The larger wild game give it a wide berth, and nothing larger than a jackrabbit will remain in a neighborhood that has so little to commend it.

Another day's travel and we pause for awhile near the headwaters of the Humboldt river. Here are two large flowing springs of equal size and about twenty feet apart, the water of one is icy cold, that of the other boiling hot.

The above-mentioned river has an extreme length of about three hundred miles, and its general course is southwesterly. It has some peculiar characteristics. At certain places along its course its fine volume of water disappears either entirely or dwindles to most insignificant proportions, and then, further along the line, it again rises in its bed and once more assumes its former size. This occurs again and again, until finally, in a great sandy waste known as the Humboldt Sink, it vanishes from view and is seen no more.

We follow along the river's banks a distance of probably two hundred miles, crossing many of its tributaries, sometimes finding an abundance of feed for the cattle, and then again being under the necessity of driving them for miles back off from our road into the smaller valleys, where good grazing could usually be found.

I have omitted to mention heretofore a custom long in practice by us. For six or eight hundred miles feed in the immediate neighborhood of our line of travel has been more or less scarce, and at times entirely lacking.

Such a state of affairs naturally created in our minds much anxiety, for on the condition of our draught animals depended the success of our undertaking; so in order that we might to a certain degree overcome this

34

difficulty, we adopted a system of selecting a certain number of men, usually two, on whom would devolve for a specified length of time the duty of finding suitable stopping places for noon and night. These men, mounted and well armed, would ride together ahead of the train, select a spot for the noon stop and await our coming, that we might be apprised of the exact locality. On arriving, the teams would at once be uncoupled from the wagons and, with their yokes on, turned out to graze. To find a good camping place for the night was a much more difficult task. We were often compelled to go three, four and even five miles out of our way before such a spot could be found. You know, other trains taking this same route west had preceded us, and the supply of feed, at no time very plentiful, had now become pretty well exhausted. If near a river, our hunters would swim across and search for grass along the farther shore; and it became a common practice with us to swim our animals over to the side opposite our camp, where, guarded by eight or ten men, they would graze all night and then swim back in the morning.

The reader of this narrative, if he follow us closely, will see that although so far our expedition has been allowed to go its way in peace, we have not in the slightest degree relaxed the stern vigilance adopted as a part of our daily life since first we crossed into the zone of danger. At night the wagons are so arranged as to form a corral, making an excellent fort in case of an attack by Indians. The method of procedure was to place the wagons one after the other, end to front, as closely as possible in a continuous line in the form of a circle, but when such circle is less than half completed, a space of about twenty feet is left vacant, and then the line is continued on beyond this gap until approaching the end, where the line was first begun, another open space like the first and opposite to it, is left. These openings or entrances are closed when necessary by stretching across them log chains securely fastened to the wagons at either side. This enclosure was useful in many ways. If any of our stock happened to be a little unruly and hard to catch, they could be driven into this arrangement and then be easily secured.

But we will resume the narrative of our journey down the Humboldt. At times the course we are pursuing brings us quite close to its banks, then again in making straight cuts from one bend of the river to another to save distance, we would be miles away. One day near its waters we came to a beautiful valley several miles in extent. It was here seven years later that a party with whom I was making my second trip across the plains to California was attacked by Indians. But of this I will speak at length further on.

As we travel on down the Humboldt the matter of securing sustenance for our stock becomes a burning question. With each succeeding day grass becomes less plentiful and our animals in consequence have lost strength so steadily that it is no longer possible to average more than ten miles per day.

We are now about two hundred miles from the source of the river and have reached the parting of the ways.

From here two separate trails lead to California, one by way of the Humboldt Sink and Carson River, known as the Carson Route, so named in honor of Kit Carson, the famous pathfinder. The other the old Oregon trail, both leading through great sandy deserts.

Consulting a guide book in the possession of our party, we are told that in taking the former course it would be necessary to pass over an expanse of deep sand forty-five miles wide, and beyond that to cross a range of very high, rugged mountains. This was the route taken three years before by the ill-fated Donner party, whose terrible experiences, when they became known, made the civilized world shudder.

After due consideration and much discussion regarding the merits and demerits of the two routes, we decided in favor of the Oregon trail — notwithstanding the disagreeable fact that directly in our pathway lay an almost waterless desert of loose, shifting sand ninety miles in breadth. But the die is cast, and so, finding near the river a spot where the supply of grass is fairly good, we lay by for three days in order that our stock may recuperate, and to carefully prepare ourselves for the ordeal. We cut and tie into bundles a large quantity of dry grass and store it in the wagons for use on the way. Finally, everything being in readiness, we set out just at noon, going in a northwesterly direction. All that afternoon and night our train crawled slowly on across this silent barren waste, and then, just as the day was breaking, we reached Rabbit Springs. Here there was only a weak flow of water, but we managed to replenish our water casks and to give to each head of stock about one gallon. Lingering here but a little while, we again pushed on, and for another day and night stopped only now and then to rest and feed the cattle.

At this time we came in sight of eight or ten giant springs, their rippling contents so clear and pellucid that small pebbles could be distinctly seen lying on the gravelly bottoms some twenty feet down. But this sparkling water, as if to mock the thirst now almost consuming us, was boiling hot.

Our cattle, almost maddened by the sight of the water, could hardly be restrained from rushing into the cauldrons, and one poor beast did fall into the hot fluid. We dragged him from it with ropes, but he was so badly scalded we in mercy killed him.

Here we were compelled to leave several wagons, the teams that drew them having entirely given out. These were unyoked and left to die. A few of the wagons were cut asunder and made into carts. Quantities of flour and bacon are abandoned and left by the roadside. We struggle on with what cattle and wagons are left, but the loose sand makes heavy pulling. The oxen stagger along at the rate of a mile an hour, frequently dropping down in their tracks completely exhausted. Allowing them to rest thus for a little while, we help them to their feet and urge them on. All of the afternoon of the last day we are in sight of Mud Lake, where good water and grass abound. If we can only get our animals that far, all will be well, but now many are so weak they cannot be urged to go farther. Men take empty kegs and, going forward on foot, bring back water to the poor famishing brutes and revive them. The stronger teams are in this way encouraged to creep on, and about sundown on the third day we are at the margin of the lake. In addition to the good water found, there is also an abundance of fine green grass. We go back and forth during the night, carrying these to the poor creatures that have so long and so faithfully served us, but only a few can be revived and brought in. Among the cattle thus lost was an old ox noted for his good sense. His owner, Sam Caldwell, had always made quite a pet of him and the two were great friends. When this faithful beast at last sank down exhausted and dying of thirst, Sam took a cup of water from his almost empty keg and attempted to revive him, at the same time calling him lovingly by name. But, too far gone to recover, the poor animal, bending his great, mournful eyes upon his master, lowed faintly in answer and expired. Sam turned sadly away, saying: "Dear old fellow, sensible to the last."

The crossing of this desert occupied over fifty hours. In the passage we lost nearly half of our wagons, many of our cattle, and were compelled to abandon a large quantity of provisions.

37

CHAPTER FIVE

E REMAIN IN CAMP at Mud Lake for rest and recuperation, now so much needed by men and beasts. The latter are very poor in flesh and weak; their strength must be husbanded, for it is a long way yet to our destination.

From Mud Lake it is only a few days' travel to the eastern base of the Sierra Nevada mountains, so leaving the Oregon trail, we go west, through a hilly country, to the foot of the Sierra range.

Compelled by the weak condition of our stock to move slowly, one whole day is consumed in reaching the summit, through the mountains, at this point, are neither high nor rough.

Down the long slope on the west side, and on through a country heavily timbered with pine trees, we find ourselves after several days' travel on the shores of Goose Lake. From here our course is south and southwest, and after traveling along Pitt River for a day or two, we at last reach the Sacramento valley. For three days we rested on the banks of the tributary known as Deer Creek and here Mr. David Myers, Sr., died of mountain fever.

The weather now became threatening and a snowstorm seemed imminent. Deer were seen in great numbers, evidently migrating from the higher to the lower altitudes for the winter. Such signs led us to believe that we had crossed the last of the great mountain ranges none too soon.

The three days' stop here was especially welcome to me inasmuch as I was just recovering from a long spell of sickness. While on the summit of the Sierra Nevadas, I became seriously ill with mountain fever. My father became much alarmed at my condition, and there being no physician with us, he sent a messenger for a Dr. Powell, who was with a train about ten miles ahead of our own. This gentleman was soon at my bedside in the wagon and, after subjecting me to a careful examination, turned to my father and said: "I do not believe it possible that his life can be saved, but I will do my best."

Giving me some treatment, he prepared to rejoin his own party, saying it was useless for him to stay with me during the night, as I could not live until morning. But my father's pleadings finally won, and he remained.

The morning came and found me better, and this improvement in my condition continuing, the doctor returned to his train; meeting him frequently afterwards in the gold diggings of California he would refer to the incident and tell those around him how he had raised me from the dead. I have often thought since, as have my father and brother, that he really did save my life. There is in my heart deep gratitude for his services, and may God bless him is my prayer.

So here in our camp on the shores of Deer Creek, though still weak and somewhat emaciated, I am convalescent.

Deer are so plentiful around us, the camp is always supplied with fresh meat. One buck killed by a man traveling with another train weighed two hundred and eighty pounds. Mr. William Hopper was the first of our party to bring in a deer. Wishing to make some soup, I asked him for a small piece. He replied: "Certainly, Billy; the sick shall have some before any one else," and he continued: "Poor old Mrs. White, who I do not think is long for this world, shall have a piece at once also." Mr. Hopper was a most kindly man. Seven years after the incident just related his daughter and I were married.

And now lowering clouds and chilling winds warn us that it is time to leave this Deer Creek country. This warning is accentuated by the discovery of certain marks, indisputable proof, on the surrounding timbers, showing that the snow here attains to a depth of from eight to ten feet. No other incentive is needed to hasten our departure. For to be snowbound now and helpless, compelled to remain long dreary months in one desolate spot, the final outcome may be a repetition of the indescribable horrors experienced by other snowbound travelers. And this, too, when almost in sight of that for which we had struggled and toiled and suffered so long to attain? Oh, no; not that, for heaven's sake, not that! And so there is hurrying to and fro, wagons are quickly put in order, cattle caught, yoked and coupled to them, whips crack and we are once more on the move down toward the valley of the Sacramento.

Soon after leaving Deer Creek, Mr. William Massey died of mountain fever. For several days now our course lay through a very rough country abounding in high hills. One day our train had arrived at the top of one of these. The road leading to the valley below was perfectly straight, very steep and perhaps four hundred yards long. It had been cleared of the small oak brush that still stood thick, like a fence on either side. So abrupt was the incline, it was thought necessary to rough-lock both hind wheels of our wagons if we would make the descent safely. One light wagon, drawn by a single yoke of oxen, drove up into position to be locked, but was halted too far forward over the edge of the declivity. The wagon, being on a slight downgrade, began crowding the oxen, and before the chain could be put through the wheels the whole outfit, minus the driver, shot away downward like a flash of light, gathering speed as it went, and was soon lost to view in a great cloud of red dust. Strange to say, the animals, poor and weak as they were, kept their feet to the very bottom, where one of the wheels, striking a big rock, caused the yoke to snap asunder, and the two principal actors in the comedy quietly turned about and began eating a lot of grass that was in the wagon, thinking evidently that the occurrence was simply a part of the regular programme, and therefore not entitled to be thought the least bit surprising.

One day more and we pass into the eastern edge of the Sacramento valley. Here there is another death from mountain fever. Jerry Overstreet is the victim and we bury him on a little hill just above the valley.

On our way down the Sacramento, we pass Lawson's place. He is the man who laid out the road we have been traveling, from Mud Lake on, the same being called "Lawson's cut-off." Our course is now steadily down the Sacramento River, along which we get our first sight of Spanish horses, and saddles with wooden stirrups.

It is now about the first of October, feed for our stock is excellent, and we see numerous bands of fat Spanish cattle roaming over the Sacramento plains.

When within twenty-five miles of the junction of the Feather and Sacramento rivers, we turn east and strike the former stream at the foothills and cross over to the eastern side, then wind through the hills for twenty or twenty-five miles further, until, on the fifteenth day of October, we reach the mining camp known as Bidwell's Bar, named after the rich bar on the Feather River discovered by John Bidwell, the same John Bidwell who, years afterward, in 1875, became a candidate for the office of Governor of California on the Independent ticket.

So the long journey is at last at an end and we are in the land of our dreams, after having been five months and eight days, or about one hundred and sixty-two days, on the road. In all that time not a word from home has reached us, neither have those there received any tidings from the father and brothers that so many months ago vanished from their sight and were swallowed up by the mists of the vast wilderness. What of the result?

On leaving home, it was confidently expected that we would return within eighteen months, at least partially successful. But it is six long years — years of labor and disappointment, of hopes and fears — ere many of us go back. Some remain in California the balance of their lives, and others still of the little band that started forth that lovely May morning, their hearts filled to overflowing with bright, glowing hopes for the future, lie under the silent stars, sleeping the sleep that knows no waking.

But here at Bidwell's Bar we write home, and John Bidwell himself, who is on his way to Sacramento, promises to mail our letters there. Many months must go by before it will be possible for an answer to reach us, but we will be patient and wait.

Our cattle are now unyoked and turned out to graze for the last time. Among the low hills grass is plentiful and they should do well, and surely theirs is a well-earned rest. But we never saw the faithful brutes again. The Indians living in the adjacent mountains ran them off and butchered them, every one. While out prospecting for gold the next summer we found their heads in an Indian village.

A few days after our arrival at Bidwell's Bar Tom Fristo was attacked with cholera morbus and died within a few hours. His was the fourth death to occur in a family of five that belonged to our party, John Kearns being the only survivor. His father, two uncles and a brother-in-law having passed away.

The diggings known as Bidwell's Bar, where we now are, covers about one acre of ground, and is of course all located. The dirt yields from ten to one hundred dollars per day to the man.

We have as yet done no good so far as mining is concerned. The precious metal can be found almost anywhere in small quantities, but under the conditions existing here the deposits must be rich to make them worth the working. Provisions of all kinds are very scarce and high in price; pork, flour and beans bringing from one dollar to one dollar and fifty cents per pound. The supplies brought by our party were now about exhausted, and most of us were out of money also.

To make matters worse the rainy season set in about the twenty-fifth of

41

October, earlier than usual, so the old timers say, hence most of the foodstuffs that had been brought in got wet before they could be properly housed. During the winter that followed we paid one dollar per pound for flour that had become almost a solid mass, and had to be cut from the barrel and the lumps pounded into a powder before it could be made into bread. Certainly a poor article of diet, but better than no bread at all.

There was much activity in the matter of building during the early winter. Cabins were constructed of shakes split from the pine trees that grew all about us. Charlie Clark paid my father twenty-five dollars per hundred for a sufficient number of these rude boards to build a house; and then ten dollars a day to help in its construction, which seemed to us better than the uncertainty of prospecting for gold.

John Bidwell employed hundreds of Indians to collect gold for him along the banks of the Feather River, giving for this service all the wheat they could eat.

I was sick during the whole of the winter of forty-nine and fifty, never having fully recovered from the spell of mountain fever previously referred to in this narrative. The disease resulted in chronic diarrhea, which finally had so weakened me that I walked with difficulty. One day while in Charlie Clark's store I saw a sack of dried blue figs, and Mr. Clark, seeing they had attracted my attention, invited me to eat some. I did so and they seemed to me delicious. Putting a few in my pocket, I returned to camp, feeling better than usual.

Very near to us at that time there was living a Dr. Clarke, a man who undoubtedly possessed great ability along the lines of his profession. He had all winter taken much interest in my case, having been at all times most kind in advising me concerning my daily diet, etc. Going to him, I asked his advice as to the figs. Becoming quite angry at the mere mention of so dangerous a proceeding he said: "You darn fool, go and eat of those figs and you will die before another sun rises." But I did not die, nor did the doctor ever hear of my indiscretion, but from that time on I continued to improve and was in time completely restored to health and strength.

In March we moved up to the South Fork and located what is called a river claim. The beds of streams were in that day supposed to be very rich in gold. It was our intention to wait until summer, when the water would be low, and then, by turning the stream from its course, expose the bed and at our leisure extract the gold from its sands.

Arrival at Bidwell's Bar

In the meantime, while waiting for the waters to subside, and dreaming day dreams of the enormous wealth that would soon flow steadily in upon us, we spent the time in prospecting. While thus engaged, we did succeed in finding a few small rich pockets in a spot, where, in after years, was located one of the richest mining camps in the State, and fortune after fortune extracted from the very ground away from which we had turned, tired and disgusted with its barrenness. This rich spot was known as the Oregon Flat.

But now the late summer had come with its low waters, and with an enthusiasm born of hope, we headed joyfully for the river claim on the South Fork. Here pitching our camp, we, with much patience and labor, at last succeeded in turning the stream from its bed and found — well, hardly a color of gold in return for our pains.

It was the decree of fate that we should not strike it rich in the diggings. But we were not the only ones thus doomed to disappointment. In our travels about the mining districts we found on an average ten men in search of the "paying" claim to one that found it.

As a business proposition, mining for gold is most uncertain in its results and demoralizing in its influences, inasmuch as it unfits a man for other and more legitimate pursuits.

So far as I am concerned, it seems to me that my lack of success in the mines was a blessing in disguise, and there is no regret on my part that matters shaped themselves as they did.

And now a word in defence of the men of '49. Although they may not need it, for the world at large has long since, I believe, recognized their true worth, representative as they were of that class of brave hearts that first carried the banners of civilization into Kentucky, Ohio and other then wild regions of our country. But from some sources malicious aspersions have been cast upon the characters of these men as a class. These attacks, however, may have been due simply to a total ignorance of the true facts in the case. I do not believe, and I say it with all sincerity, that it would be possible to find in all the world a more orderly, honest, sober set of men than these miners. During my stay of twelve months among them, I did not see one single fight, and but two or three cases of drunkenness. There was absolutely no thievery. Although beyond the jurisdiction of the law they were still law-abiding, and in their dealings with each other they were governed by a spirit of fairness and justice that was most commendable. And all this, too, while ar away from the tender, restraining influences of the home and of society.

The famous Sam Jones once in a letter to a friend said he found Sacramento to be one of the roughest and toughest towns he had yet preached in in California, and that he was told that this was to be accounted for by the fact that a large percentage of the citizens were forty-niners, and he added that he "thanked Heaven that within a few years all of the forty-niners would be dead."

I at once lost faith in Sam, knowing he was expressing views on a subject he know nothing about.

In September Amos Kusic, a negro, and a sailor who had been his mining partner, having between them about three thousand dollars, left Bidwell's at noon on their way East. After going about three miles they sat down on a divide to rest. Here they were attacked by Joaquin Murietta and his gang of outlaws. One of the men was lassoed and dragged to death, while the other was killed with a knife. When the bodies were discovered there was found a trail of blood leading away from the scene of the crime. It would seem that the negro had in some manner during the encounter come into possession of a knife belonging to one of the robbers, and had used it with deadly effect.

A posse was organized and followed the Spaniards for many miles, but never succeeded in overtaking them.

45

CHAPTER SIX

IN NOVEMBER, 1850, my father and I, being desirous of seeing more of the country, left the mines and came down into the valleys. Reaching Sacramento, we purchased a packhorse and other things necessary for the journey contemplated.

On the third day of December we crossed the river on a ferryboat and headed west.

The first day after leaving Sacramento we arrived on the spot where Davisville now stands. Jerome Davis had settled there that fall. The second day saw us a little below the present site of Winters; and on the third day out, which was December 6, 1850, we reached the valley which bears our name. The whole country was at that time filled with wild game. Hundreds of elk could be seen in a single herd, and antelope were equally numerous, while great flocks of wild geese covered thousands of acres of ground at a time. Deer were very plentiful and quite tame. Of these as many as a hundred or more could be seen in half a day's hunt. And I must not forget to mention that royal beast, the monarch of them all, the grizzly bear. This region was his home, and for years after my father had settled in this valley he continued to challenge our right to oust him from it.

Two miles above what is now the town of Winters we found living on the south side of Putah Creek John R. Wolfskill. He had settled here in 1841. He gave us much information concerning the country and the little handful of people in it. Where Vacaville is now there was one small house, owned and occupied by a Mrs. McGuire. Two of the Longs also lived in that neighborhood. In Lagoon Valley there were two Spanish families, named, respectively, Barker and Panier. These were the only settlers within twenty miles of our home.

During the summer of 1851 Richardson Long, Whit Long, M.R. Miller, Henry Ammens, George Egbert and William Smith all settled in this valley. In 1852 John Simpson made his home here. G.W. Thistle came in 1857. So about all of the Government land was now taken up. The years immediately following 1850 were busy ones with us. Virgin forests that had never, since the world began, heard the sound of the axe had to be cleared away, buildings erected and the ground put in condition for cultivation.

Those who have never had any experience in such matters can hardly realize the enormous difficulties to be overcome in founding a new home in a wilderness. The task of a new settler is a hard one, even when within easy reach of stores and sawmills, but in a country where these bases of supplies are either entirely lacking or are so distant as to be almost out of the question, the problem is doubly perplexing. But we managed it somehow, and within a few years were well established.

In 1856 my father sent me to Missouri to bring the remainder of our family to California.

February the 19th, I left home bound for San Francisco, from which place it was my intention to proceed to St. Louis, Mo., by way of Panama and New Orleans. My ticket, calling for a steerage passage, cost me one hundred dollars to the latter city. On February 20th I sailed for Panama on the ship "Sonora" and reached there in about twelve days, after an uneventful voyage. The railroad that carried me across the isthmus to the Atlantic side was the first I had ever seen, but, having read much about them, it was about as my mind had pictured it.

After crossing the isthmus I went on board a large vessel, named the "George Law," bound for Cuba, and reached the city of Havana in due time. From there the ship "Queen City" carried me to New Orleans in just twenty-two days from San Francisco.

The journey from New Orleans to St. Louis was made on a steamer, the name of which I have forgotten. We were nine days on the trip, which cost sixteen dollars. At St. Louis I took the train for St. Charles, a distance of twenty miles or so, and as far as the North Missouri railroad was at that time completed. Stopping over night in St. Charles, I purchased the next morning a horse, saddle and bridle for one hundred and sixty dollars and rode the animal two hundred miles or more, to the home of my brother and sisters in Western Missouri, stopping on the way, however, at a point fifty miles west of St. Louis for a week's visit to my uncles Edward and Royal Pleasants.

It was on the sixth day of April that I arrived at the home of my brother and sisters, just forty-seven days from the time I left my father's house in California. The trip had, altogether, been a most enjoyable one.

The only traveling I had ever done before was the crossing of the plains in '49, and soon after reaching California then I had settled down and remained amid the peaceful environments of our valley home, living a life of innocent freedom and content. Excessively fond of hunting, no one ever had finer opportunities than I of indulging in a favorite sport, so numerous were grizzly bears, elk, antelope, deer and panthers, or California lions, as some call them. So this journey East was a new and interesting experience to me, especially that portion of it made by water. There were no storms or bad weather of any kind encountered, and after the first day out I was not bothered by seasickness.

It was exceedingly pleasant to be again in the company of my relatives, after an absence of six years. After my visit to my uncle, Edward Pleasants, was ended and I was taking leave of him he said: "William, be a man and remember you have to die." I have never forgotten and never shall forget those words and his impressive manner as he uttered them.

After visiting with my brother and sisters for a few days, during which I gave them a history of all that had transpired since we had last met, I began making preparations for our journey to California. A party intending to cross the plains was then being organized in the neighborhood. I at once joined the company and began the purchase of an outfit. Here was a good opportunity to get rid of some of the gold that I had carried constantly on my person since the day I left California. While in San Francisco I had bought a buckskin money jacket that fitted my body next the skin snugly. This garment had in it numerous small pockets, and in these I carried all the money I possessed, a sum amounting to about fifteen hundred dollars in gold coin. For nearly two months I had worn it constantly. At first it did not inconvenience me much, but the longer it was worn the heavier it seemed to become and at last when it was discarded my skin bore for weeks afterward visible impressions of ten and twenty-dollar gold pieces. In fact, I was so impressed with gold money at that time that the impression of its usefulness has lingered with me ever since. It may seem strange to the reader of these pages that I should carry this money so long in the manner stated, but you must remember that in those days there were no banks, and as to leaving it with friends; well, I knew myself better than I knew them and preferred to be my own treasurer.

Within fifteen days I had secured for five persons, these being my three sisters, my brother and myself, the following outfit: Two saddle horses, four choice milk cows, six yoke of good oxen and one wagon. Besides these there was a sufficient supply of clothing and enough provisions to last us six or eight months, everything bought being substantial and useful. This outfit, in some particulars, differed from those of the others, inasmuch as I, profiting by my former experience, laid in more dried fruits and sugar and less bacon.

Many of the older men thought that I, being only a boy, was making a mistake. But these afterward realized that they were the ones who had been mistaken. By May the sixth, 1856, we were ready to go. The start was to be made from Big Creek, a point on the open prairie in Johnson County, Missouri. Soon all had arrived at this spot. There were altogether eight wagons and about fifty men, women and children in the party. Two of the men, Messrs. William Hopper and David Burris, were old comrades of mine, they having been members of the company that I crossed the plains with in '49.

At the hour of starting friends and relatives had gathered from far and near to bid us good-by, and such sorrow and weeping I never witnessed before. The scene was a touching one. The parting of these people, many of them never to meet again on earth, was affecting in the extreme. But the leave-taking was at last ended, wagons were lined up, the word to start give and I was again on my way to my California home. We traveled about ten miles that day, and, on camping for the night, we proceeded to elect a leader of the expedition, and Mr. William Hopper was unanimously chosen as captain of the train. No other officers were deemed necessary.

I was now at the zenith of my delight, being headed for my beloved California home. There were now no special ties binding me to the East. My brother and sisters being with me, there remained in that section no member of my immediate family; hence, unlike the others, it was with a light heart and joyful anticipations that I turned my face toward the West. My life on the Pacific Coast had well fitted me for the present journey. Healthy and strong and exceedingly fond of the chase, I longed for the day to come when we would be among the buffaloes and antelope out on the great plains.

But how different were the feelings of the others. They were leaving home and dear ones to face for months unknown hardships and dangers, and finally to dwell among strangers in a strange land. For a few days there was much despondency and gloom among them, but this gradually wore off amid the new and interesting scenes through which we were passing. We about us the prairie grass was abundant, while the weather continued beautiful. It was in this portion of the country between the Missouri line and were soon across the Missouri line and in the wild country that afterwards became the State of Kansas.

We crossed the Kaw or Kansas River a few miles west of where the town of Lawrence was afterwards located, our general course being northwest. All the Big Blue River that three of our party got lost one night in a dense fog. It was customary for us to have three men always on guard with the cattle at night, the first watch lasting from dark to midnight. The second watch would then relieve the first and remain on duty until day. On the night in question the fog had settled down so thickly that it was possible to see but a short distance away. The cattle, as was usual at night, had been rounded up in a small space about five or six hundred yards from the wagons. The second watch relieved the first at midnight, and in so doing showed them what direction to take in order to reach the wagons. The camp fires had gone out and, the people being asleep, there was neither light nor noise to further direct the men to the right spot, and in the dense darkness they missed their way and were soon wandering here and there, totally lost. Their absence from the camp was not discovered until after daylight, when the second watch came off duty. Then measures were at once taken to find them. Guns were fired at intervals to apprise them of our whereabouts should they be in hearing distance, and finally Uncle Charlie Hopper jumped on his mule and set out, at the risk of being lost himself, to search for the missing ones.

About five miles away he came across their trail in the dew on the grass and, following this, overtook them six or seven miles from camp. The names of these three men were William Hopper, Ike Isley and John Sackett. Our train now averaged about twelve miles per day; and we were soon among the buffaloes and antelope, which resulted in an abundance of fresh meat for all.

One morning on the banks of the Little Blue River we had halted our train in order that the cattle might rest and graze for a few hours when a party of about one hundred Indian warriors belonging to the Cheyenne tribe rode boldly into our camp. But as luck would have it, over one-half of the men in the company were at the time preparing for a target shoot and, to the surprise of the savages, were ready, with their guns in their hands, for any emergency. As it turned out, however, not a shot was fired, but no telling how the meeting might have ended had there not just at this time come in sight a large force of men convoying a Government train. These Indians were armed with guns and lances, and each also carried on his arm a shield. The heads or blades of the lances were made of steel about one and one-half inches in width and three and one-half feet in length, having very sharp edges, tapering each way from the center. Their points were very sharp. These blades were set into wooden shafts one and one-half inches in diameter and the whole, head and all, about fourteen feet long. The things looked to be more dangerous than six-shooters, and it is said the Indians use them with great skill and accuracy, and that in a close fight they are more dangerous in their hands than rifles would be.

The officials with the Government convoy, with whom we afterwards talked, told us that this band undoubtedly intended to plunder our train, but, finding us prepared for a fight, somewhat disarranged their plans and then the Government train happening along settled the matter in our favor. Anyway, they stood not upon the order of their going but went at once, and we saw them no more.

CHAPTER SEVEN

UR ROUTE WAS NOW UP THE Little Blue River for several days. It was on the banks of this beautiful stream in 1849 that the expedition I was then with, buried its well-loved wagon master, John Lane. Now, in company with Mr. William Hopper, I went to the place where we had laid him just six years before. All that remained at this time to mark the spot were a few fragments of the head and foot boards that we had erected above the grave. It was with feelings of the greatest sadness that we turned away from this lone last resting place of noble John Lane. From the Little Blue we turned northwest through a rolling hill country, and within two or three days were on the south bank of the Platte River near Fort Kearney. This stream is described in a previous portion of this narrative.

The valley of the Platte is the favorite home of the buffalo; and, for the sake of the younger generation that may some day happen to read this little volume, I will describe to the best of my ability one of the many buffalo hunts in which I was an active participant. Early one bright sunny morning, just as our people were breaking camp preliminary to a continuance of the daily journey westward, David Burris, William Wester and myself, well armed and mounted, left the busy scene and rode up a small stream, a tributary of the South Platte, known as Plum Creek, headed for the table-land some distance away, where we might view the surrounding country and choose the ground for the day's sport, for at this time the buffalo were to be found in such vast numbers that the hunter could at his leisure, and according to his own fancy, select the location that was to be the scene of his operations, just as school boys choose a spot for playing ball. Reaching the elevation that had been our objective point from the start, there was spread out before us a scene that will forever be denied the future generations of men. In that clear transparent atmosphere, east, south and west, as far as the most perfect eye

52

The Buffalo Chase

could reach, were to be seen countless thousands of buffaloes, not huddled up together in bunches but separated from each other, say from three to five to the acre, quietly grazing like so many cattle. We rode up quite close to where a few were rolling in the dirt, as we sometimes see horses roll, a practice common to the buffalo. We dismounted and, each man selecting an animal as a mark, fired, but seemingly, without effect. Mr. Wester then proposed that one of us, suggesting himself, take charge of the rifles, the other two to ride among the animals armed only with Colts revolvers. This plan was at once adopted. With a revolver in one hand and the bridle in the other, Burris and I put spurs to our horses and dashed towards the quarry. Singling out one fine animal as the object of attack, we both made for him. Our horses seemed to partake of the spirit of the sport and bore us nobly. It was our intention to ride up close to the great beast until our horses' necks were on a line with his hips and then shoot him in the upper flank. This was not an easy thing to do, for when our horses would get close enough to catch the odor of the buffalo they would, in spite of us, shy to the left and run away, but after several attempts and as many runaways we succeeded in approaching near enough to get in some telling shots, which made the beast furious. At every shot he would wheel and charge us, and more than once he came near ripping the side of a horse with his sharp, curving black horns. These thrusts so frightened our horses that it was as much as we could do to keep our saddles. The pursued animal finally stopped running and stood at bay, madly pawing the ground, first with one foot and then the other, his blood-shot eyes full of rage and defiance. Standing with lowered head in a position of defense, he seemed to dare us to come on.

It is a waste of ammunition to shoot a buffalo in the face. The front of the skull, naturally of great thickness, is rendered more impervious by a long-standing accumulation of sand and dirt several inches thick matted in the woolly front. So our only chance was to take running shots at his side. This we did and finally dispatched him.

We were now some ten or twelve miles south from the emigrant road, and a little line of blue smoke curling upward from a grove of quaking asps at the head of Plum Creek, only two or three miles away warned us that we were in close proximity to an Indian camp, and we thought it best to turn about and as soon as possible get to where our guns were, far back to the rear in the hands of Billy Wester. So when we had each secured a nice chunk of meat for ourselves and an extra good piece for our comrade we started back and were soon all together again. After riding hard the balance of the day we overtook our company just before dark, after they had gone into camp for the night. This ended one of the most exciting hunts I ever engaged in.

It was at this time that I began to understand and appreciate the extreme bitterness with which the Indian regarded the encroachments of the whites, realizing, as he no doubt did, that no matter in what manner or how long he might resist, the superior intelligence of his foe must in the end conquer. Forced further and further back, away from the beloved lands where the bones of his fathers mouldered, he could see that it was only a question of time when his proud race would be doomed to extinction. The wild game that now nourished him and his loved ones must soon disappear before the numberless rifles of his masters. Want, physical decay and disease must follow, and in the end a race of men worthy of a better fate will have perished from the face of the earth and not one left to tell of the tragedy. Ah, well, 'tis the old, old story being repeated, and that has been repeated over and over again, since first the world began, the survival of the fittest, and man's inhumanity to man.

In this, as in many other portions of the country, we often saw and were deceived by that strange phenomena known as a mirage. The tired traveler riding for hours, maybe a whole day, in the stifling heat of the desert without having tasted a drop of water and almost famishing from thirst, sees suddenly in the distance ahead a beautiful lake. Mirrored in its bosom are to be seen the clear-cut outlines of the far-off curving hills, while on its lovely banks groves of trees cast their cool shadows over grassy swards and bend low their luxuriant foliage to kiss the sparkling waters. Almost maddened by the sight, man and beast hurry forward with renewed energy over the desolate hot sands, their one thought to reach as quickly as possible the restful shade of the oasis. But the vision, with cruel mockery, keeps always the same distance ahead and finally fades from view altogether.

The mirage is usually seen across a level plain on a hot day. I have often seen them between me and a distant range of hills or mountains, these seemingly suspended in the air with the lake underneath them. Of course we soon became acquainted with this deception, as we did with some others that were along somewhat different lines. The Indians, for instance, would sometimes try to decoy us within range of their arrows by going into the tall grass, which would conceal their bodies, and then place upon their own heads the head and antlers of an antelope, at the same time counterfeiting

55

the movements of that animal. We never shot an Indian that was playing this little game, but, nevertheless, he was taking some tall chances.

The coyote takes the place of the rooster on the plains, for he invariably howls at the dawn of day. One will start the concert, another immediately follows, and soon in every direction as far as the ear can reach there is one grand swelling chorus of yells, snapping of teeth and mournful howls. Bands of wolves would gather around our camp, await our departure and then come in and feast off the refuse that had been left.

Beside the coyote there were three other kinds, all large animals, one gray, one cream-colored and another black. Although plentiful, some being in sight nearly all the time, we were never molested by them. While mentioning this wild music of the plains, I am reminded of the fact that there were among us some that played well on the violin, two of these instruments being in our outfit. There were also some good voices, and often at evening we would gather in one tent and have music and dancing, never failing, however, to keep at such times a vigilant watch, that we might not be surprised by the prowling savages.

We continued up along the banks of the Platte for a few days and then crossed to the other side, and in so doing left behind us the great buffalo country. From the north side of the river our course lay through a country abounding in low hills. After some days journeying through these we reached the North Platte, and for several weeks this stream was followed, and along its course we encountered some peculiar natural rock formations. One of these, known as Courthouse Rock, resembles in its outlines a great building, such as a courthouse or State capitol structure. Another, known as Chimney Rock, rises to a great height and is somewhat like a chimney or smoke stack. I attempted to climb to its top but could not do so by at least one hundred feet.

A few miles below Fort Laramie we passed near a spot where are buried one hundred and twenty U.S. soldiers, killed one year before in a battle with Sioux Indians. This battle was precipitated in the following manner: The Indians, numbering several hundred, were camped on the banks of the river. One or more of them had stolen and butchered an ox belonging to an emigrant train. Complaint was made to the authorities at the fort by the owner of the stolen animal, and a young officer with little or no experience in Indian warfare was sent with a company of soldiers to arrest the thief or thieves, but these the Indians stubbornly refused to deliver up. Thinking to frighten them into acceding to the demand, the young officer ordered that a volley be fired over their heads. This, on being done, so enraged the savages that they immediately attacked the soldiers and, being greatly superior in numbers, succeeded in killing nearly all the whites before assistance from the fort arrived. The poor victims of the massacre were all buried in one long grave.

Upon our arrival at Fort Laramie we found camped in the neighborhood about one thousand Indians. They had assembled for the purpose of entering into a treaty of peace with Uncle Sam. General Harney had soundly thrashed them for the massacre of the soldiers and they were now desirous of coming to terms, though still in a sullen mood.

It was some time in July when we left this point, and we continued up the river until one hundred miles or so above the fort before we crossed over. Feed is getting scarce and we are often compelled to camp several miles from the main road in order to find grass for the cattle that are now becoming poor and consequently losing strength. We have to frequently lay by to let them rest and recuperate.

The nights are growing colder, owing to our near approach to the Rockies but the days continue warm and pleasant.

Independence Rock is now met with and some few miles west of it we strike Sweetwater Creek or river, up which we travel for about three weeks. This charming little stream, coming down from the summit of the Rocky Mountains, runs due east and empties into the North Platte. Its length is in the neighborhood of three hundred miles, and we followed it for nearly that distance to its source.

The South Pass, where we had now arrived, is a gap several miles wide in the mountain range. About one-half of our journey was now completed, but our draught animals were in a bad condition, being footsore and poor. And there was ahead of us at least one wide desert to cross, where neither water nor feed was to be found. Thinking of these difficulties in our pathway, we began the descent of the Western slope and reached Green River, one of the upper branches of the Colorado. All the way to Bear River there was good grass. At a point midway between old Fort Hall and Great Salt Lake we turn away from Bear River, for this stream now bears toward the Southwest.

For a week or ten days we are among low hills, valleys and plains, then Goose Creek is reached. In this section of country there is an abundant supply of good water and grass, but the Indians, partly under the influence of the Mormons, and aided and encouraged by them, were very troublesome. For two weeks or more we posted a double guard over our stock at night, and on more than one occasion Indians or white men (we could not tell which), were seen in the darkness prowling around and were fired upon. We believed these extra precautions saved us from disaster. In this neighborhood we saw several new graves, near which were posted written notices reading: "Killed by Indians. Be careful."

It was while in this dangerous territory that I one day received a mysterious warning. As was my custom, I, one afternoon, rode on alone ahead of the train to locate a suitable camping ground for the night. After going several miles I became quite drowsy and dismounted to rest a spell. Dropping the picket rope in order that my mule might graze, I stretched myself upon the ground and was soon sound asleep. Just how long I had slumbered I do not know, but certainly not very long, when I was suddenly aroused by a human voice calling my name in a low wary tone. It said, "William, William, William!" There was no mistake. Three times I heard it clearly and distinctly, and it came from only a few steps away. In an instant I was on my feet and, fully realizing now the danger of my position, I lost no time in remounting and hurrying away. It was a strange occurrence, and I believed then, and do now, that at the time some imminent danger threatened me and the voice was a friendly warning from some mysterious unseen source.

CHAPTER EIGHT

FTER LEAVING Goose Creek our road lay through an uninteresting region. Passing Thousand Springs Valley, we came to the Humboldt River, down which we traveled its entire length, and it was on the banks of this stream that we had our most serious encounter with Indians. It was the 24th day of August, 1856. We had spent the noon hour at the upper end of a beautiful valley that was four or five miles long and about one and a half miles wide. At 1 o'clock we had hitched up and were continuing our journey when we came to another wagon train in camp on the bank of the river. We had met this same party before and were well acquainted with them. In the conversation that ensued they informed us that there was a band of hostile Indians numbering a hundred or so some distance down the valley whose actions boded no good to any one going that way, and they deemed it decidedly dangerous for us to proceed further in that direction that afternoon. In pursuance of their friendly advice, we halted and went into camp a few hundred yards away from them. We had hardly gotten our cattle unyoked and turned out to graze when it was discovered that the Indians were trying to set fire to the long grass and at the same time they began shooting the stock of our neighbors, which were grazing half a mile or so from the wagons. The grass was too green to burn, and the Indians that were shooting the animals were soon driven away by the boys on guard, not, however, until they had succeeded in killing several head of cattle and a mule. One painted warrior during this time seemed to turn his entire attention to us, dashing back and forth in front of our wagons at a distance of about one hundred and fifty yards. He was well mounted and carried a good looking rifle. Some of our party wanted to shoot him, which could easily have been done, as he presented a shining mark, but the older men forbade them to do so. At first we watched his maneuvers closely, wondering what his intentions could be,

but as he was doing us no harm we ceased paying any attention to him. Finally, riding into a cluster of willows, he fired into a group of our boys, wounding two of them badly but not fatally. The bullet passed through William Hopper's thigh and then into Harvey Pleasants' groin, where it lodged and remained until his death, which occurred just twelve years afterward.

The savage, seeing the success of his murderous work, immediately whirled his horse and rode away with the speed of the wind, and, although one of our boys fired at him, he escaped and rejoined his band. Doctor Mathews' train was in camp down the valley one mile from ours. Upon leaving us, the Indians made a dash for his outfit, took him and his party by surprise, stampeded and carried off fifteen head of horses, which was all he had, and were soon out of range of any ordinary gun. But the Doctor had with him something that was rare in those days. This was a fine long-range telescope rifle. Deeply incensed at the manner in which he had been treated, he rested this weapon upon the wheel of a wagon, took careful aim, fired and killed one of the Indians at a distance of several hundred yards.

The battle being now over, the Doctor came up and dressed the hurts of our wounded. But we were compelled to stay where we were until the boys were able to be moved, which was a week or ten days. The second night after the Indian attack two of our men followed the direction taken by them and located their camp, which was in a small valley some fifteen or twenty miles north, so one hour after dark on the following night fifteen of us, well armed and mounted on our best horses and mules, set out with the intention of surprising them but they somehow learned of our approach, and, extinguishing their fires, they came very near surprising us, for we turned back just in time to avoid being riddled by Indian bullets. They tried to decoy us back by building big fires, but we prefered to warm ourselves by blazes kindled in our own camp, and so we returned to our friends.

This was probably a marauding band of Oregon Indians led by white men, maybe Mormons, that had been preying on the emigrant trains all the season, as we afterwards heard of their depredations both behind and in front of us. However, we were never again molested by such foes.

At Gravelly Ford, a point about one hundred miles below where the Indian fight occurred, we were met by my father, who had come from California with several saddle and pack horses to meet and assist us in crossing the mountains and deserts, and right glad, too were we to see him

The Indian Attack on the Humboldt

Meeting My Father on the Bank of the Humboldt

and the sleek fat animals he brought with him. It had taken him some six weeks to make the journey from our home to this point and he had made the trip alone. It was a daring venture for a man in those days to travel a distance of more than five hundred miles without companions, crossing as he did the snowy ranges of the Sierra Nevada Mountains, and the forty-five miles of a barren, waterless desert.

The hardships and privations that he endured and suffered during his lonely pilgrimage were a touching proof of the depth of his love for, and devotion to, his children, while the addition of the fresh horses so sorely needed to help lessen the burdens now borne by our well-nigh worn out and exhausted stock, was a great blessing to the whole company and was much appreciated by all.

A short distance from where my father met us the trail forks, one road turning to the northwest. This is the old Oregon trail, over which I traveled in 1849. Our course now was down the Humboldt to that desolate waste known as the Humboldt Sink. Into this great sandy basin flow the Humboldt, Carson, Truckee, Walker and many smaller streams, all vanishing from the sight of man as their waters sink beneath the sands. The desert in which this basin is situated begins about one hundred miles north of the "Sink" and extends to the Mohave desert on the south, a distance of several hundred miles and is from fifty to one hundred miles in width. At the mouth of the Humboldt there are several hundred acres covered with bulrushes and tule grasses, and here we found the lowest and most degraded type of human beings I have ever seen. Absolutely naked, they presented a most revolting appearance. I saw them eating the raw flesh of ducks, their lips covered with blood and down, and the scene was disgusting in the extreme. I did not learn to what tribe they belonged and it is doubtful if they knew themselves.

At this point we laid by for a few days to rest the oxen and make preparations for crossing the forty-five miles of barren waterless desert on the edge of which we were then camped. With enough cooked food to last us two days, full water kegs, and wagons stored with cut grass of poor quality, but better than none, we started just at noon. The sand was deep and loose and our progress in consequence very slow, but all the afternoon we kept going.

At nightfall there was a halt of half an hour or so to rest and feed the cattle and then we pushed on again. Every one that could walk did so, in order to save the strength of our animals. At daylight there was another short interval of rest. The sun rose red as blood and soon it became exceedingly

Crossing the Humboldt Desert

not. One and one-half miles an hour was the best we could do. North, south, east or west, not a spear of vegetation could be seen. Only a great sea of hot sand on every side, the very abomination of desolation, and we the only living things in it. At noon another stop for a little while. Our train was scattered now, a distance of several miles separating the first and last wagons in the line. Both men and beasts were in great distress, many of the latter giving out altogether, but we kept on as best we could with the remaining ones. The last five miles was the worst, but at last just at sunset, after thirty hours of almost constant travel the stronger teams stood on the banks of the Carson River. I had noticed that the last ten miles of the road was strewn with the carcasses of cattle, thousands of them lying where they had fallen and all about these the wreckage of wagons bore mute testimony to the fact that others beside ourselves had suffered.

Resting a day or two we followed up the bank of Carson River for several days and reached the foot of the Sierra Nevada Mountains. The stream now led us up through a canyon where the great rocks and crags hung over us, their pinnacles so high above our heads that one almost had to look twice to see their tops. The road became so steep and rough that I cannot describe it. One can judge of the difficulties we encountered in going over it when I say it took us nearly a whole day to travel three miles. But at last we were at the first summit and rested for the night in Hope Valley. The next day we crossed the second summit, an undertaking that gave us much trouble, as we were compelled to hitch twenty yoke of oxen to each wagon, one at a time, in order to get them over. But with patient toil and care all were finally on the other side. What might have been a serious accident occurred just on the top of the second summit. One of the little boys in the party, Jeff Hopper by name, fell from one of the wagons and a wheel passed over his head and shoulder. Fortunately the wagon was a light one and had at the time little or no load on it and the little fellow suffered no serious injury.

The atmosphere at this great altitude was delicious and invigorating and my heart leaped with joy as I beheld in the dim distance the Coast Range Mountains and realized that just back of them lay my home.

Many glaciers were to be seen round about us in the heads of canyons, some of them covering many acres of ground and perhaps from fifty to one hundred feet deep. It was all wonderfully interesting to me.

What a change in the topography of the country! A few days ago we were in the midst of a great waterless sandy desert, where no living thing, either vegetable or animal, could be seen. Here beautiful flowers, babbling streams and magnificent forests were all about us.

From now on my story is a simple one. Within ten days our train stood on the banks of the then beautiful Sacramento River. No need of weary quests for suitable camping grounds now, or burning deserts to cross, no dreary guard duty in drenching rains or midnight Indian alarms. But peace and quiet and rest under the hazy October sunlight in God's country.

Two days more and we were at home, after five months and six days of travel.

[THE END.]

THE TRAIN OF 1849

Names of people who crossed the plains in the Pleasant Hill train in 1849 which started from Pleasant Hill, a small town in Cass County, Missouri:

John Green
Mr. White
Mrs. White
Miss Lizzie White
Bale Hicklin
Mrs. Bale Hicklin
Thomas More
James Kusick
Amos Kusick (Black)
Sam Kusick (Black)
Isack Sparks
Mat Sparks
Henry Sparks
Hervey Sparks
Richard Sparks
James Fleming
David Fleming
James Freeman
Jerry Overstreet
Hardin Overstreet
George Overstreet
Mr. Peacock
Mr. McKenna
Harrison Williams
William Hopper
Oren Durby
Mr. Rector
John Burris
David Burris
James M. Pleasants
J.E. Pleasants
W.J. Pleasants
Mr. Lyons
Mr. Eaton, Sr.
Mr. Eaton, Jr.
Mr. Sinclair

William Hensley
Mid Story
Emanuel (Black)
John Kearns
Mr. Kearns
David Rice
William Massy
Tom Fristo
James Hamilton, Sr.
James Hamilton, Jr.
Med. Rollen
Henry Lawrence
Johial C. Williams
James Williams
James Preston, Sr.
James Preston, Jr.
Sid Arnett
John Arnett (Black)
Mr. Idson
Robert Sloan
Old Uncle Dick Sloan (Black)
Dr. McReynolds
David Myers
John Myers
Colonel Hann
John Brisco
Charles Brisco
John Lane
Julious Right
Pret Manyon
James Coldwell
Sam Coldwell
William Parker, Sr. (Preacher)
William Parker, Jr.
Greenbery Parker
Tom Clayton

William Miller
James Allen
Andy Allen
Sam Whiteman
Mr. Walker
George Noel
Mr. Hatch
Mrs. Hatch
Miss Dora Hatch
Miss Hatch
Miss Hatch
John Hatch
James Keton

THE TRAIN OF 1856

The names of men and boys over ten years of age who crossed the plains with th writer in 1856:

William Hopper and family
Charles Hopper and family
Benton Hopper
William Hopper
Edward Hopper
James Hopper
Jeff Hopper
Mack Hopper
Columbus Hopper
John Bingham and family
David Burris and sister
Ike Isley
Matt Isley
Tom Burgin
Marion Davidson (now of Ukiah, Cal.)
John B. Plummer (of Kentucky)

Thomas Harvey Pleasants
W.J. Pleasants and three sisters
Robert H. Rhea and family
Talford Powell
John Houx
Robert Cavett
William H. Wester
Buck Barker
John Helm
John Farmer
Rolly Farmer
Henry Burris
Red John Tacket
John Tacket
George Foster and family, now o
    Fremont, Solano Co., Cal.

Index

COLOPHON

The William J. Pleasants overland to California is one more scarce to rare account of early travel to the far west in search of riches in the gold fields of California, and is one of several such overland accunts that has been printed in the workshop of Glen Adams, which is located in the sleepy country village of Fairfield, Washington, in southern Spokane County and not far from the Idaho line. The book was set in type by Miss Bobi Pearson, using a Compugraphic 48 computer photosetter. The setting was in twelve point Baskerville roman, with running heads in ten point Baskerville Bold caps, and page numbers in twelve point Baskerville Bold. The pen and ink illustrations used in the Pleasants book were in the original printing but for the Ye Galleon edition were redrawn by Helen Staples of Latah, Washington. Camera-darkroom work was by Evelyn Foote Clausen. The film was stripped into flats by Robert LaTendresse. Indexing was by Edward J. Kowrach. General book design was by Glen Adams. The sheets were printed by Robert LaTendresse using a 770CD Hamada offset press. The sheets were folded by Evelyn Foote Clausen using a multi-sheet Pitney Bowes folder. Some collating-assembly work was done by Edward J. Kowrach. Binding is by William Bosch of Oakesdale, Washington. This was a fun project. We had no special difficulty with the work.